# A DELPHINID (MAMMALIA, CETACEA) FROM THE MIOCENE OF PALOS VERDES HILLS, CALIFORNIA

# A DELPHINID (MAMMALIA, CETACEA) FROM THE MIOCENE OF PALOS VERDES HILLS, CALIFORNIA

BY

LESLIE E. WILSON

UNIVERSITY OF CALIFORNIA PRESS
BERKELEY • LOS ANGELES • LONDON
1973

UNIVERSITY OF CALIFORNIA PUBLICATIONS IN GEOLOGICAL SCIENCES

Volume 103

Approved for publication April 14, 1972
Issued April 15, 1973

UNIVERSITY OF CALIFORNIA PRESS
BERKELEY AND LOS ANGELES
CALIFORNIA

UNIVERSITY OF CALIFORNIA PRESS, LTD.
LONDON, ENGLAND

ISBN: 0-520-09458-1
LIBRARY OF CONGRESS CATALOG CARD NO.: 73-83101

PRINTED IN THE UNITED STATES OF AMERICA

# CONTENTS

# ABBREVIATIONS

| | |
|---|---|
| C.A.S. | California Academy of Sciences |
| U.C.M.P. | University of California Museum of Paleontology |
| U.S.N.M. | United States National Museum |
| U.C.M.V.Z. | University of California Museum of Vertebrate Zoology |

# A DELPHINID (MAMMALIA, CETACEA) FROM THE MIOCENE OF PALOS VERDES HILLS[1], CALIFORNIA

BY

LESLIE E. WILSON

(A contribution of the University of California Museum of Paleontology, sponsored in part by the Annie M. Alexander Endowment)

## ABSTRACT

A NEW SPECIES of a Middle Miocene genus of delphinid, *Loxolithax* Kellogg, Delphinidae, Phocaeninae, is represented by a badly crushed skull (including tympanic bullae, periotic bones, and auditory ossicles), a left forelimb lacking phalanges, 29 vertebrae and several ribs. These fossils were discovered in the upper part of the Valmonte Diatomite Member of the Monterey Shale (late Micene in age) in a quarry in the Palos Verdes Hills, Los Angeles County, California, 33° 47′ 30″ north latitude and 118° 21′ 00″ west longitude. The new species differs from the type species, *Loxolithax sinuosa* Kellogg, from the Sharktooth Hill bone bed, Kern County, California, in detailed features of the periotic bone, especially in the more dorsal position of the *fenestra rotunda,* the wedge-shape of the *pars cochlearis,* and the sharp-crested elevation of its dorsal surface above the adjacent area containing the dorsal orifices.

## INTRODUCTION

THE BONES OF this new fossil delphinid were discovered by McKinley Stockton, superintendent of the Dicalite (or Dacelite) Company's 1936 quarry, about three miles south of Lomita, California. The specimen was taken to Dr. Loye H. Miller, then at the University of California at Los Angeles. Dr. Miller communicated with Dr. C. L. Camp of the Museum of Paleontology of the University of California at Berkeley, who arranged to have the writer transport the specimen from Los Angeles to Berkeley for study.

Cetaceans travel far and fast and therefore should furnish good paleontological means for correlating marine sediments over long distances, even from one continent to another. Because of this wide-ranging habit among cetaceans, it might be supposed that students of the fossil whales would be inclined to be lumpers rather than splitters in their taxonomic nomenclature. The number of fossil whales in museum collections have been inadequate to permit determination of variation within a species. Examples of over-split taxa in both fossil and living cetaceans are readily found in the literature.

In the early days of American cetacean studies (summarized by Kellogg [1925]; R. Harlan [1824]; J. Leidy [1851]; E. D. Cope [1868], and other writers) it was the practice to give generic and specific names to fragmentary remains of cetacean fossils. The genera *Priscodelphinus, Ixacanthus, Delphinapterus, Tretosphys,* and *Belosphys* were based almost exclusively upon vertebrae. In Europe, Viscount Du Bus (1867; 1872) discovered cetacean material in the Antwerp basin consisting of skulls, fragments of skulls, and a few vertebrae. In his 1872 paper he established

[1] The United States Board of Geographical Names has adopted the name Palos Verdes Hills for this peninsula, which was formerly known as San Pedro Hills. See decisions rendered between July 1, 1934 and June 30, 1935, p. 19, 1936.

seven species on fragments of skulls, two species on two crania, and one species on "many" skulls, cervical vertebrae and some anterior dorsal vertebrae. He then assigned these ten species to Cope's genus *Priscodelphinus,* which had been established on an imperfect posterior dorsal vertebra. When using fragments of skulls or only vertebrae, which are variable, the opportunities for setting up new genera and species were almost unlimited. The theoretical intermingling of cetacean fossils from different sides of the Atlantic did not clarify the problem of proper assignment of the fossils to genera and species. In extenuation of the faults in procedure of some of the earlier paleontologists it may be said that the lack of adequate material restricted their ideas on variation, and that the millstone of special creation still hung about their thoughts. If a species had been specially created its members would be identical whether found in Maryland or Belgium or anywhere else.

It remained for Case (1904) and True (1908; 1912) to discover in the Miocene of Maryland fossil cetaceans having both skulls and skeletons associated, and representing a few of the same genera found in Europe. While awaiting discovery and description of more specimens, it seems advisable to establish new species or genera on differences in the bones of the ear, on changes in the arrangement of the bones that form the apex of the cranium, in the degree of telescoping of skull bones, or in the number, arrangement and morphology of the teeth. Kellogg (1931) had only periotic ear bones of the small odontocetes from Sharktooth Hill near Bakersfield, California, upon which to base his generic and specific names. Until delphinid skulls with earbones in place are found, other fossils from the North Pacific area must be compared with Kellogg's taxa only on the basis of the periotic bones.

## ACKNOWLEDGMENTS

Thanks are expressed to all of the members of the professional and technical staff of the University of California Museum of Paleontology at Berkeley for their personal interest and encouragement. To those with other affiliations, thanks are given to the late Dr. Loye H. Miller for presenting the specimen, to the late Dr. Remington Kellogg of the United States National Museum for his scholarly counsel, and to Dr. Charles A. Fleming, senior paleontologist of the New Zealand Geological Survey, and to Dr. R. A. Falla, director of the Dominican Museum, Wellington, for making available the skull of *Kekenodon onamata* Hector to the late Dr. R. A. Stirton who kindly made a sketch of the stapes of this Oligocene cetacean for the writer. Excellent photographs of the stapes and periotic of *K. onamata* were made by Mr. F. O'Leary of the Dominican Museum. All drawings are by Mr. Owen J. Poe, staff artist of the University of California Museum of Paleontology. The delicate ear bones were prepared by my wife, the late Edith Wilson.

## SYSTEMATICS

Order CETACEA Brisson, 1762

Suborder ODONTOCETI Flower, 1867

Family Delphinidae Gray, 1821

### Genus *Loxolithax* Kellogg, 1931

The genotypic species, *L. sinuosa,* was based on an isolated left periotic, C.A.S. 4352 (Kellogg, 1931: figs. 127, 128), and a paratype left periotic, C.A.S. 4351 (Kellogg, 1931: figs. 129, 130) both from the Sharktooth Hill bone bed, Round Mountain Silt = "Temblor" Stage = Middle Miocene. Our new material makes possible a more comprehensive generic diagnosis. Inasmuch as no skeletal parts except the periotic of this fossil were previously known, an extended diagnosis is given herewith.

*Revised diagnosis*—Cranium short, broad, and deep; brain case probably inflated; frontals small, contacting supraoccipital in a narrow smooth crest; nasals long and narrow, thickened posteriorly and forming the highest point of the cranium; mesethmoid thick, forming a conspicuous ridge between the nares; premaxillae widest in front of the narial openings, forming the premaxillary eminences; inner borders of premaxillae cover the anterior part of nares; premaxillaries edentulous; extend posteriorly to points opposite the middle of external narial openings, with blunt posterior terminations; foramina of premaxillae are posterior to those of the maxillae. Maxillae comprise the larger part of the dorsal surface of the skull; posteriorly they spread outward and contact the supraoccipital along the greater part of its anterior edge; broadly expanded behind the orbit, overriding the supraorbital processes of the frontals but not completely covering the lacrimals; bordering the nares the internal edge of each maxilla forms a smooth rounded fold; maxillary notch of moderate depth, formed by the lacrimal and maxilla; anterior maxillary foramen is on a transverse line with the maxillary notch; rostrum broad at its base, tapering rapidly to its blunt anterior tip; it makes up little more than half the length of the skull, a ratio approaching that of the Recent *Phocaena;* supraoccipital slopes gradually in a posteroventral direction from its contact with the frontals; the dorsal part of the supraoccipital shows no concavity behind the contact with the frontals as is the case with the Eurhinodelphidae, which suggests that the braincase was probably inflated; cervicals free; posterior and anterior processes of periotic angular; anterior process laterally compressed; ventral face of posterior process slightly concave, broad, and ridged; large fossa surrounding dorsal orifice of *aqueductus vestibuli;* dorsal surface of noncochlear portion of periotic strongly convex and not flattened on dorsalmost part.

## *Loxolithax stocktoni*[3], n. sp.

(Figs. 1-5, 6*e-f*, 7*c-d*, and 8)

*Holotype*—U.C.M.P. 34576, a mature skull with tympanic bullae, petrosals and auditory ossicles, mandibles, 29 vertebrae, a left forelimb lacking phalanges, and several ribs.

*Type locality*—U.C.M.P. locality V-36118, 33° 47′ 30″ north latitude and 118° 21′ 00″ west longitude. Dicalite (Dacelite) Company Quarry, including open pits and shafts for excavation of diatomite on the northeast edge of the Palos Verdes Hills and south adjacent to the community of Walteria, Los Angeles County, California.

*Horizon*—The upper part of the Valmonte Diatomite member of the Monterey Shale at the Dicalite Quarry from which the new species of *Loxolithax* was taken, is considered by Woodring (1946) to be Upper Miocene in age, based on foraminifera identified by R. M. Kleinpell. Dr. W. W. Wornardt, Jr. (pers. commun.) examined a sample of diatomite taken from the block containing the skeleton described herewith, and confirmed the Valmonte Diatomite as the provenance of the fossil. He suggested a Late Miocene age, "probably Mohnian." Lohman (1965) contributed the information that his abundant collections of diatoms from the Calvert Formation, Middle Miocene, of Chesapeake Bay "form a very distinctive assemblage which is almost identical with that found in the middle Miocene Temblor Formation of California, in particular, with the 'Indicator bed' in the Temblor Formation on Anticline Ridge near Coalinga and with the middle Miocene Round Mountain silt on Sharktooth Hill near Bakersfield." Kellogg's type periotic bone, on which the genus *Loxolithax* was established, came from the Sharktooth Hill bone bed, which is part of the Round Mountain Silt. Thus it appears that there is reasonably good correlation between the Calvert Formation of Maryland, which has yielded many fossil dolphins, and the Sharktooth Hill bone bed from which the type of *Loxolithax* came. The Valmonte Diatomite is not as yet correlated paleontologically with either the horizon of the type of periotic or with the Calvert Formation of Maryland. For the present the biostratigraphic position of the new species is designated as Upper Miocene.

---

[3] The species is named in honor of McKinley Stockton, who discovered the specimen.

## Diagnosis

The new species differs from the type in the following ways: the *pars cochlearis* shows in the cerebral view a wedge-shaped dorsal surface appearing as a ridge which forms the external boundary of the internal acoustic meatus. The corresponding area in the type is more greatly rounded; the posterior internal groove in the species is deeper and wider than in the type; the ventral face of the posterior process of the type is more sharply concave than that of the new species; the area between the prominence above the fossa for the head of the malleus and the posterior process, which area articulates with the sigmoid process of the tympanic bulla, is more irregular in the new species than in the type; the relatively large size and triangular shape of the *fenestra rotunda* in the new species differs from the type, as detailed later. Stapes, *fenestra rotunda*, and the dorsal orifice of the *aqueductus cochleae* lie in a straight line.

## Description

### SKULL

The posterior third of the sutures on the skull between the maxillae and premaxillae are fused and in part indistinguishable, hence a mature animal is represented. The cranium is short, broad and deep with an expanded facial portion and occipital shield. The rostrum is bluntly terminated and represents little more than one-half the total length of the skull. The specimen is crushed, causing the occipital region to be pressed into the brain cavity and the lower jaws and their teeth to be pushed against the maxillae throughout their entire length. Proportions of the skull are like those of the living porpoise, *Phocaena*. The rostrum is blunt, being broad at its origin and tapers rapidly to the anterior tip. The length of the symphysis relative to the length of the mandible in *Loxolithax stocktoni* is short, 23 percent. In *Kentriodon pernix* Kellogg it is 32 percent, in *Eurhinodelphis bossi* Kellogg, a long-beaked dolphin, 55 percent, and in a living *Delphinus* (U.C.M.P. 23071), 17.5 percent. Thus, in this character, *L. stocktoni* resembles the phocaenines and short-beaked delphinines and is separated from *Kentriodon pernix*, which it otherwise resembles in many ways.

*Dorsal view*—The highest point of the cranium is formed by the nasals, which are relatively long and narrow, and thickened posteriorly. The right nasal is more completely preserved because crushing has driven the left nasal forward where its anterior end lies below the mesethmoid.

The frontals are small and apparently fused. Posteriorly, they join the supraoccipital in a narrow smooth crest. The mesethmoid is complete. It is exceptionally thick posteriorly where it is high enough to form a ridge 8.0[2] thick between the nares. The premaxillae are narrower than the maxillae and are thin bones from their anterior tips to a point opposite the anterior maxillary foramina, where they increase in width and thickness, reaching their greatest width just in front of the narial openings. The inner borders of the premaxillae form part of the borders of the narial openings. Because of the fusion of the premaxillary-maxillary sutures, it is difficult to determine the posterior limits of the premaxillae. After careful search, in which use was made of various liquids on the bones in order to heighten contrast, the

[2] All measurements are in millimeters.

conclusion was reached that the posterior termination of the premaxillae is slightly posterior of the middle of the external narial openings. At its posterior end the left premaxilla has a width of 8.0, and the right, 11.5. The premaxillary foramina are small and slitlike and are located just posterior to the maxillary foramina. The premaxillae are in contact with each other from the anterior borders of the narial openings to a point approximately 40.0 from their anterior tips, at which point they diverge slightly.

The maxillae comprise the larger part of the dorsal surface of the skull. Anterior to the maxillary notches they taper rapidly to a point near the ends of the premaxillae. The maxillae override the supraorbital processes of the frontals but do not completely cover the lacrimals. Posteriorly they flare out and contact the supraoccipital along the greater part of its anterior edge. Anterior to the maxillary notches the outer borders curve gradually downward to the tooth row. The maxillary notches are of moderate depth and the adjoining lacrimal is plainly delineated on the left side. The anterior maxillary foramen on the relatively undistorted right side is on a transverse line with the maxillary notch. On the lateral border of the narial opening, the internal part of each maxilla forms a smooth, rounded ridge, which is formed by the curling outward of the internal edge, so that the bone overhangs itself along a length of 30 mm. This overhang is better preserved on the left side than on the right.

*Posterior view*—The supraoccipital slopes gradually in a posteroventral direction from its contact with the frontals. Ventral to a transverse break in the bone located 17.0 below the contact with the frontals, the posterior part of the braincase has been moved forward into the brain cavity. The occipital condyles pass into the supraoccipital dorsally and into the exoccipitals laterally with a smooth contact and no suggestion of a neck. The outline of the foramen magnum suggests a pentagon, with sharp and straight lateral inferior borders. Before crushing, the supraoccipital probably bulged posteriorly as in the modern phocaenines. The sturdy paroccipital process of the exoccipital probably extended more than 37.0 below the ventral edge of the occipital condyle.

*Ventral view*—The ventral surface of the rostrum is smooth and broadly convex, its ventral-most portion being, except in the anterior one-half, along the midline. The maxillary part of the ventral surface of the rostrum is interrupted only by part of the palatine vacuity of the right side. The pterygoids encompass large and transversely expanded sinuses. The part of the basicranium that is floored by basisphenoid and basioccipital widens posteriorly, and its inferior borders are formed by the descednding falcate process of the basioccipital and the internal plate of the pterygoid.

The zygoma is shaped like a broad leaf curled at its edges. Its right side has a maximum length of 55.0 and a maximum width of 31.0, the left side is 56.0 and 29.0. The glenoid fossa is circular in outline, only slightly concave, and has a diameter of 18.0. It occupies a much smaller part of the zygoma than in the living *Delphinus.* The squamosal is incomplete on both sides of the skull.

The left maxilla contains 28 teeth and three empty alveoli. The right side shows a jumble of 29 teeth in a distance of 102.0, some of which may be from the crushed right mandible.

The jugal has a long, thin styliform process, the width of which varies from 3.2

near the posterior end to 2.8 near the anterior end. It is broader than the styliform process of the modern *Delphinus*. The total length of the jugal is 74.5. The boundary between lacrimal and the supraorbital process of the frontal is clearly visible at the external contact.

*Lateral view*—In lateral view the cranium is high, the paroccipital process is well developed, the zygoma thickens anteriorly attaining a thickness of 18.0 near its anterior end, and is strongly arched throughout its length. The supraorbital process of the frontal is strongly arched and has a long posterior extension, and the rostrum is shallow and smoothly tapering anteriorly.

Height of the cranium from the posterior end of the basioccipital to the superior surface of the nasals (slightly oblique) is 97.0, but may have been decreased by distortion during crushing. External edges of the maxilla and of the supraorbital process of the frontal are coextensive except at the anterior end of the process, where the maxilla does not override. The anterior part of the supraorbital process has a thickness of 9.0. Its posterior extension descends ventrally and is stout throughout. It is similar in shape to that of *Kentriodon pernix*. The anteroposterior dimension of the dorsal part of the supraorbital process measures 27.0.

### Measurements of the Skull

| | |
|---|---|
| Greatest length of the skull | 314.0 |
| Length of rostrum (sagitally, maxillary notch to anterior extremity) | 184.0 |
| Rostrum: skull length index | 58.0 |
| Distance from maxillary notch to posterior edge of paroccipital process (right side) | 148.0 |
| One-half width of rostrum at maxillary notch (right side) | 45.0 |
| One-half width of skull at anterior edge of nares | 76.0 |
| Distance across premaxillae at anterior border of nares | 50.5 |
| Width of right premaxilla at anterior maxillary foramen | 16.5 |
| Greatest width of right premaxilla at anterior edge of nares | 26.8 |
| Least distance between inner margins of premaxillae on vertex of cranium | 28.0 |
| Greatest anteroposterior diameter of right nasal along suture | 26.5 |
| Transverse diameter of right nasal at anterior end | 15.0 |
| Transverse diameter of right nasal at posterior end | 17.5 |
| Maximum transverse dimension across nasals | 34.4 |
| Least distance on midline from posterior edge of nasals to anterior edge of supraoccipital | 20.5 |
| Width of exposed portion of left frontal on vertex | 22.5 |
| Height of cranium (ventral surface of falcate process of basioccipital to dorsal surface of nasal bone) | 100.0 |
| Estimated height of foramen magnum | 30.0 |
| Estimated width of foramen magnum | 30.0 |
| Greatest distance across occipital condyles | 76.3 |
| Estimated height of right condyle | 40.0 |
| Greatest height of left condyle | 40.0 |
| Greatest width of right condyle | 29.5 |
| Greatest width of left condyle | 24.4 |
| Greatest length of right zygomatic process (ventral side) | 55.0 |
| Greatest length of left zygomatic process (ventral side) | 56.0 |
| Length of left jugal | 74.5 |
| Greatest thickness of maxilla and frontal combined on outer margin near center of orbit | 9.7 |
| Length of right lower jaw | 270.0 |
| Length of symphysis | 62.0 |
| Symphysis: mandible length index | 23.0 |

## TEETH

The teeth of the left maxilla are the most easily accessible for study. In the lower jaws and the right maxilla many teeth are preserved, but because of the crushing of the mandible against the maxilla, the arrangement of teeth cannot be ascertained with certainty. In the left maxilla 27 teeth are in place, of which 9 are complete. The others have been broken off at or below the junction of the crown and root. In addition to 27 teeth, there are three empty alveoli, two at the extreme posterior end of the maxilla, and one near the center. The anterior ends of both premaxillae are lacking and it is possible that two or three more teeth were present in *L. stocktoni,* or the premaxillae may have been edentulous as in *Delphinavus* "San Luis Obispo County, California, Vaqueros?," and the living *Delphinus* or *Phocaena.* Thus the number of teeth in the upper jaws of *L. stocktoni* is at least 30, and not more than 32. The incompleteness of the premaxillae in *L. stocktoni* is particularly unfortunate, for it is in this region that definitive comparison would be made with *Kentriodon pernix* "Miocene, Calvert Formation, Maryland." In *K. pernix, fide* Kellogg (1927:33), the extremity of the premaxilla bears an enlarged anterior tooth that projects directly forward, with only slight downward inclination.

The number of maxillary teeth in *L. stocktoni* separates this taxon from the genus *Kentriodon,* which show 40–40, also from *Delphinavus,* which shows 40–40, and from *Sinanodelphis izumidaensis* Makiyama, "early Miocene, Bessyo Formation, Nagano-ken, Japan," which apparently has 45–45. Also the teeth of *Kentriodon* are about twice as large as those of *Loxolithax.* Thus, in number of maxillary teeth, *L. stocktoni* is close to Recent *Phocaena.* Because of the crushing, it is not possible to estimate the number of teeth in the mandible of *L. stocktoni.*

The posterior teeth are larger than the anterior, although there is not a regular gradation. The crowns may be either larger or smaller than the roots, are often flattened in a lateral direction, and curved lingually at the tips. Frequently, the neck of a tooth is constricted, giving the appearance of the head-end of a match. This shape is common in the mandibular teeth. Others taper gradually to a sharp point, the apex curving in a lingual direction. Anteroposterior flattening may be seen in several teeth. Longitudinal marks, apparently from wear, were observed on several teeth in which the tip of the crown is well worn.

There seems to be no regular arrangement of the different kinds of teeth. The fourth tooth, counting forward from the alveolus at the posterior end of the left maxilla, has a rugose tip and possibly an accessory cusp. It might be expected from a study of *Delphinodon dividum* True (1912:172) that the teeth with accessory cusps would be at the posterior end of the series. The average length of teeth of *Delphinodon* in True's specimen is 29.0, and in *D. leidyi* they reach a length of 38.0, but in *Loxolithax stocktoni* the largest tooth found in only 15.5 in length. The interalveolar septa are so thin that the teeth almost touch each other. The teeth are hollow and the roots open, making them fragile, and the crowns of most are lacking.

One of the teeth (fig. 8*e*) pressed against the right maxilla, in the tooth row but with its root not in place, has three accessory cusps below the apex of the crown in addition to the central cone. Several other teeth show suggestions of an extra tubercle. One tooth has a longitudinal groove. The tooth with the accessory tubercle has an enlarged root, and a constricted neck just below the chown. The height of its crown is 3.8, and its greatest diameter is 3.0. The diameter of the neck is but 2.6 and below it the root enlarges to 3.0. The length of the longest tooth found is 15.5 and its greatest diameter is 3.5. The length of its crown is 5.0.

Comparative Sizes of Teeth

| | *Total length* | *Greatest diameter* | *Length of crown* |
|---|---|---|---|
| Largest tooth found | 15.5 | 3.5 | 5.0 |
| Smallest tooth found | 7.5 | 2.0 | 2.5 |

## EAR BONES

Within a family of cetaceans the bones of the ear region are remarkably similar in major outline. Hence, reliance on morphologies of these bones as indicators of phyletic relationship appears to be well founded.

At least as early as 1858, Claudius recognized the value of the structure of the internal ear as an aid to the classification of animals. Subsequently other investigators; Hyrtl, Ibsen, Denker, and especially Fraser (1883) and Doran (1879); established the possibility of identification of the genera of living mammals on the basis of the ear bones, and recognized the value that may be attached to the bones (ossicles) of the middle ear. Brandt (1871), in the absence of other skeletal elements, established a new genus of squalodont on the basis of a tympanic bone. Kellogg (1931:375–397) proposed seven new genera and nine new species of the family Delphinidae in a collection of 24 periotic bones from the Sharktooth Hill bone bed. The validity of these genera will be tested in the future in two ways: the establishing of limits of variation by the study of many periotics of the same species of living delphinids; or even better, the discovering of fossil ear bones associated with other critical skeletal parts, as is described in this paper. In the present state of knowledge of the history of the odontocetes, examples of transitional stages in evolution are few, too few to recognize phyltic lineages.

Among the characters of the periotic bones used by Kellogg (1931:359–361, 367–371), for specific differentiation, the shape of the *pars cochlearis*, presence or absence of a crest on its tympanic face, topographic configuration of the area between the *aqueductus vestibuli* and the posterior rim of the acoustic meatus, type of crest (if any) posterior to the stapedial fossa, and shape of the extremity of the anterior process have proven to be useful in the present study. Variations of less than specific value were considered by Kellogg to be: the position of the *foramen singulare*, the size of the *aqueductus vestibuli*, and some changes in the internal acoustic meatus. Six specimens of periotics demonstrating variation in these last-named characters were all assigned to the same genus and species, i.e., *Lamprolithax simulans* Kellogg.

Preparation of the skull of *Loxolithax stocktoni* revealed the presence of both left and right ear bones in place. Because of the crushing of the skull, it was decided to remove only the elements of the right side for detailed description. The following table of measurements compares the periotics of *L. stocktoni* with homologues in *L. sinuosa*.

MEASUREMENTS OF PERIOTIC BONES IN LOXOLITHAX

| | *L. stocktoni* U.C.M.P. 34576 Type | *L. sinuosa* C.A.S. 4352 Type | *L. sinuosa* C.A.S. 4351 Paratype |
|---|---|---|---|
| Greatest length (tip of anterior process to tip of posterior process) | 31.6 | 29.6 | 28.0 |
| Width in line of *fenestra ovalis* (from external face above excavation to internal face of *pars cochlearis*) | 18.5 | 16.8 | 16.3 |
| Greatest dorsoventral depth of *pars cochlearis* (from most inflated part of tympanic face to most salient point on cerebral face) | 10.5 | 10.0 | 10.5 |
| Distance between nearest edge of *fenestra rotunda* and tip of anterior process | 19.4 | 18.1 | 17.8 |
| Distance between nearest edge of *fenestra rotunda* and tip of posterior process | 18.4 | 15.8 | 14.9 |
| Distance between nearest edge of ventral orifice of *aqueductus fallopii* and tip of anterior process | 14.5 | 14.2 | 14.2 |
| Greatest length of posterior process | 14.5 | 13.5 | — |
| Greatest width of posterior process | 10.9 | 9.4 | — |
| Ratio of width of posterior process to its length | 75% | 70% | — |
| Greatest diameter of *fenestra rotunda* | 2.1 | — | — |
| Distance from nearest edge of internal acoustic meatus to nearest edge of *fenestra rotunda* | 3.0 | — | — |
| Distance between external aperture of *aqueductus cochleae* and *fenestra rotunda* | 4.5 | 4.2 | — |
| Width of excavation between the posterior process and the anterior tuberosity | 2.2 | — | — |

*Right periotic* (figs. 6*e-f*)—In ventral (tympanic) view the outstanding characters are: the length and breadth of the posterior process, the depth and size of the groove for the facial nerve, the size and depth of the fossa for the head of the malleus, the lateral compression of the anterior process, and the wedge-shaped *pars cochlearis.*

The posterior process is large in comparison to the anterior process. The external rim of the anterior half of the process is raised above the rest of the process into a sharp ridge; the elevated anterior edge overhangs the canal for the facial nerve. A broad central longitudinal groove is depressed between two prominent osseous ridges which diverge from the middle of the anterior edge toward the tip; the anterointernal rim of the process is elevated less than the external. This arrangement of features gives a basinlike appearance to the ventral face of the process; the posterointernal boundary is a straight line.

The type of *L. sinuosa* (figs. 6*c-d*) differs in the following details: The face of the posterior process of the type periotic is more basinlike, especially between the two prominent radiating osseous ridges than the new species; the posterior end of the facial canal is partially walled off by an elevated rim of bone much higher than in the new species; anterior edge of the type more elevated than in the new species. There are two pits at this end of the canal; a straight, narrow incisure borders external side of the posterior process which deepens in an anterior direction. In the new species the comparable groove is much wider. The type has a slightly larger fossa for the head of the malleus than the new species and the surrounding rim of the bone has a sharper area between the posterior process and the anterior eminence which is much shorter and more angular in the type than in the comparable area in the new species. The anterior eminence in the new species is lower and more rounded than in the type. The openings in the acoustic meatus of the new species are larger than in the type, and the bony partitions separating the openings are longer and narrower in the new species than those of the type.

The canal for the facial nerve, or the *aqueductus fallopii,* is broad and deep and partly roofed on the outer side by a projecting ledge of the ventral face of the posterior process. Opposite the stapes, the canal divides into two parts, the more dorsal division passing into the ventral or epitympanic orifice (facial foramen) of the facial canal. The canal, beyond this orifice, passes through the periotic in a dorsomesial (cerebral) direction and emerges on the dorsal side, with its opening near the base of the anterior process. The more ventral and smaller division ends in a deep fossa beside the ventral orifice and the *fossa incudis;* the latter receives the *crus bevis* of the incus.

At the posterior end of the facial canal, there is a large, partly enclosed fossa for the stapedial muscle. The internal border of this fossa extends down the side of the *pars cochlearis,* where it forms the border of a thin ridge on a level with the ventral external side of the *pars cochlearis.*

The *fenestra ovalis,* closed by the foot of the stapes which is in place in this specimen, is slightly elliptical in outline, with a major diameter of 2.3 in an anteroposterior direction.

The fossa for the head of the malleus is approximately circular in outline and sufficiently deep to be marked by steep internal sides. Its diameter from rim to rim in an anteroposterior direction is 3.0. It lies at the base of the prominence just behind the anterior process. The anterior process is laterally compressed and has a blunt anterior end. Its general direction of curvature is inward, and its dorsal side slopes in a ventral direction. The *fenestra rotunda* is barely visible on the posterior side of the *pars cochlearis* in ventral view. The outline of the *fenestra rotunda* is that of a spherical triangle of about 2.0 on each side. The distance from its nearest corner to its closest approach to the facial canal is 3.7. The stapes, *fenestra rotunda* and the dorsal orifice of the *aqueductus cochleae* are in line.

In dorsal view, the periotic of *L. stocktoni* shows five foramina, with bony partitions between them. The posterior process has a rugosity lying at the ventroexternal edge of the entrance to the facial canal. The whole process curves sharply in an external direction, and its surface is irregular at its external extremity. The external aperture of the *aqueductus cochleae* is a minute pit located on the upper posterior rim of the *pars cochlearis.* In this specimen it could not be completely explored.

The external aperture of the *aqueductus vestibuli* is located just external to the *aqueductus cochleae* and is a slitlike aperture extending in a nearly externointernal direction and located at the bottom of a depression, which begins at the apex of the ridge connecting the posterior and anterior processes. The dorsal (cerebral) aperture of the facial canal is located anteriormost of all the foramina. It is a large, elongate and deep slit about one-half the size of the internal acoustic meatus. The direction of the long axis of the slit is oblique toward the external side of the anterior process, crossing the anterior process at an angle of thirty degrees. The depression surrounding the *foramen singulare* is small and round and separated from the dorsal aperture of the facial canal by a bony partition which is noticeably wider than the diameter of the *foramen singulare.* This foramen is located posterior to the dorsal aperture of the facial canal.

A bony partition separates the *foramen singulare* from the rest of the internal acoustic meatus. The fossa for the internal acoutic meatus is the largest and most internal of any of the openings in the dorsal view. It is separated at the bottom by bony partitions into four depressions, the most anterior being the spiral tract ending in the *foramen centrale.* The posterior division of this fossa shows a continuation of the spiral tract of the anterior division. The spiral, after passing upward in the anterior division, crosses above the bony part and passes downward to the bottom of the posterior depression, exhibiting here a chamberlike structure, the openings of which may be mistaken for minute foramina unless the bottom of the chamber is well lighted.

*Right tympanic* (figs. 7*c-d*)—The tympanic bulla was only loosely attached to the periotic. Its thin outer lip is broken away throughout the greater part of the length of the bone, exposing part of the *cavum epitympanicum.* Numerous fragments of the lip are crushed into the tympanic cavity. The sigmoid process curves from the lip of the tympanic toward its point of contact with the periotic. Its posterointernal face is flattened as shown in the table of measurements. The *involucrum* shows little variation in height throughout its length, the maximum height being 9.1 distant from the anterior tip. Its internal face is nearly straight and slopes slightly in a ventroexternal direction. This face is rugose. The dorsal face is marked by faint ridges radiating from its central external edge toward its internal edge. The ventral face of the bulla is divided by a deep groove extending the length of the bulla. This groove is deeply incised at the posterior end and becomes broader and more shallow in an anterior direction. At the center of the bulla the groove curves toward the external edge of the bulla but shortly returns to the central line.

MEASUREMENTS OF THE RIGHT TYMPANIC

| | |
|---|---|
| Length of tympanic from external posterior tip to anterior tip (posterior pedicle broken off and measured separately below) | 33.6 |
| Maximum width of tympanic just anterior to sigmoid process | 19.5 |
| Width of extreme posterior end | 16.0 |
| Depth from ventroexternal corner to top of sigmoid process | 19.1 |
| Maximum depth of involucrum on internal side (dorsal surface of involucrum to ventral surface of same) | 11.6 |
| Greatest depth of tympanic groove at posterior end | 7.1 |
| Diameter of sigmoid process at top of arc | 3.6 |
| Distance across arc of sigmoid process from one end of sigmoid process from one end of sigmoid process to other end | 8.2 |
| Width of flattened area of sigmoid process on internoposterior side | 3.1 |

*Posterior pedicle of bulla.*—This process is a crudely wedge-shaped bone, with the greatest thickness at the posterior end. It was attached to the dorsoposterior end of the tympanic by a narrow contact. Its ventral face shows several pits encircled by a rim of bone for contact with the tympanic. Its dorsal face is ridged and grooved to fit the ventral face of the posterior process of the periotic in a free articulation.

| | |
|---|---|
| Greatest anteroposterior length of pedicle | 16.1 |
| Greatest width at center of pedicle | 12.4 |
| Greatest dorsoventral thickness near posterior end | 8.4 |

*Right malleus* (figs. 8*a-b*)—The malleus is complete except for the extremity of the tip of the large anterior process. It consists of two principal divisions: first, the head and extension of the bone to which the head is attached; and second, the long process, the posterior end of whose base attaches to the external side of the smaller facet of the head, and the anterior end of its base make up the anterior end of the whole malleus. Thus, the base of the process is practically the same length as the whole malleus. Inasmuch as the ossicles of the middle ear were found in place, their orientation can be given with accuracy. A line through the center of the anterior process makes an agle of 50° with a line connecting the extremities of the anterior and posterior processes of the periotic on the anterior side. The point of the process is directed externoanteriorly and ventrally. The body of the malleus is oriented with its long axis making an angle of about 50° on the posterior side with the line connecting the extremities of the anterior and posterior processes of the periotic.

The head of the malleus is divided into two facets for articulation with the incus. These facets are nearly at right angles to each other and have as a common line of contact a narrow groove, wider at its anterior than its posterior end. These facets are plainly delineated from the head by small grooves. The dorsal side of the head is ovoid and extends slightly beyond the facets in a posterior direction. Attached to the external side of the smaller (more ventral) facet is the posterior end of the base of the large anterior process of the malleus. This process curves over from the external to the internal side, leaving a foramen between it and the smaller facet for articulation with the incus. This foramen passes through the malleus and probably served as the passage of the *chorda tympani* nerve. The extension of the protuberance anterior to the head of the malleus has no definite neck. The manubrium extends externally from the base of a large process near the anterior end of the base of the malleus. On the dorsal side of the anterior process is a deep groove which follows in a posterior direction and terminates in the above-mentioned foramen at the anterior angle of the junction of the facets for articulation with the incus.

In general in Odontoceti, the manubrium is globular in form (Fraser and Purvis, 1960:117, fig. 29, pl. 49) and bears a small pointed process identified by Hyrtl (1845) as the short process, to which the apex of the *tensor tympani* muscle is attached. The mechanics of vibration, transfer and amplification from the tympanic membrane to the foot of the stapes is clearly set forth by Fraser and Purves (1960: 129 *et seq.*). The anterior process (= *processus gracilis* = *processus longus* = *processus folianus*) lies adjacent to the sigmoid process of the tympanic bulla and fuses with the bulla (Kellogg, 1927) in the groove between the sigmoid process and the uncinate process. In *Loxolithax stocktoni* no evidence of fusion of the anterior process of the malleus to the tympanic bulla is clearly indicated.

MEASUREMENTS OF THE MALLEUS

| | |
|---|---|
| Maximum median length of large process of malleus | 4.6 |
| Maximum width at base of large process of malleus | 6.0 |
| Maximum length from posterior end of head to anterior tip of extension in line with the head of the malleus | 6.6 |
| Maximum diameter of head of malleus | 2.9 |

*Right incus* (figs. 8*c-d*)—The termination of the *crus breve* of the incus appears to be complete; and if so, the incus is complete. Its facets and *crus longum* articulate with the malleus and the stapes respectively. The facets of the incus match in size and shape their articulating facets on the malleus. The larger facet of the incus is nearly flat but the smaller is deeply concave. The *crus longum* forms the larger part of the incus, is sufficiently large to coincide with the boundaries of both facets, is strongly arched in a posterior direction, and bears a deep pit on its anterior side near the articular facets. Its posterior end bears a small facet, elliptical in outline, for articulation with the stapes. The *crus breve* is a minute process about 2.0 in length and 1.0 in diameter, arising near the border of the larger of the articular facets. It appears to be too short to articulate with any part of the periotic and may have had cartilaginous connection.

MEASUREMENT OF THE INCUS

| | |
|---|---|
| Greatest diameter across articular facets | 2.6 |
| Distance from end of *crus longum* to base of large facet | 2.9 |
| Distance from face of large facet to *crus breve* | 1.0 |

*Right stapes*—The stapes, a tapering column with an elliptical cross section, is firmly in place, closing the *fenestra ovalis.* It flares outward at its lower end into an annulus which covers about one-half its length, the diameter of the lower one-half being twice that of the upper one-half. Just above the annulus the column is pierced by a circular opening, the intercrural aperture, extending in a transverse direction and passing through the column in the plane of the short diameter of the ellipse. This aperture is for passage of the stapedial artery, a branch of the internal carotid (Goodrich, 1958: fig. 490, p. 464).

MEASUREMENTS OF STAPES

| | *L. stocktoni* | *Lagenorhynchus obliquidens* (L. Wilson collection) |
|---|---|---|
| Overall length of stapes (vertical) | 2.4 | 2.4 |
| Distance from top of head to intercrural opening | 1.0 | 1.5 |
| Anteroposterior length of head | 1.5 | 1.5 |
| Internoexternal width of head | 1.1 | 1.0 |
| Anteroposterior length of base | 2.3 | 2.4 |
| Internoexternal length of base | 2.0 | 1.5 |

The comparative measurements given above suggest there has been little or no change in the stapes of delphinids of similar size since Late Miocene time. If this stability is generally true, the stapes has little value in determining relationships in post-Miocene delphinids.

Fraser (1882:446) wrote: "The cells of the embryonic stapes appear contemporaneously with those forming the embryonic cartilage in the arches, or those forming the periotic capsule; they are arranged in a circular form round an artery, which may either disappear very early, as in the human embryo, or persist through life, as in the rat. In the former case it is called arteria stapedialis, in the latter arteria stapedio-maxillaris. This circular ring is at first of equal thickness all round, and is not even in contact with the periotic capsule, but is more closely connected with the hyoidean cartilage. . . ." Fraser also found that in parts of the stapes, the head and margins of the base are the last to appear. It appears from this work of Fraser that the stapes never was columnar.

In his extensive work on the comparative anatomy of whales, Slijper (1936:58–65) tabulates the findings of many authors on the embryology of whales. In general, he finds that the *fenestra ovalis* is closed shortly after birth by a membrane that grows against the margin of the foramen and the foramen is entirely closed by the time the individual reaches sexual maturity. Slijper (1936: fig. 42, p. 85) shows a sagittal section of a 48.0 cm. embryo of *Phocaena phocaena* in which the stapedial artery lies between the two well-separated shanks of the stape s. At this early stage of development there is obviously no single post for the stapes. Slijper noted that at this stage the course of the stapedial artery was not easy to follow beyond the stapes. His investigations show (p. 85) that in young cetaceans the intercural opening is still clearly defined, but in adult specimens the opening is completely or nearly closed. Exceptions are the stapes of *Loxolithax stocktoni* and the stapes of a Recent dolphin *(Lagenorhynchus obliquidens).*

The close similarity between the Miocene and Recent form also aroused my curiosity about the possible change in shape and size of the intercural aperture during the course of development of whales. A search for the record of size of the intercural aperture in the archaeocete whales showed that *Dorudon osiris* (Dames) of the Upper Eocene, Fayum, Egypt, has an oval intercural aperture (Kellogg, 1936:193), but its size is not given. *Kekenodon onamata* Hector shows an opening for the stapedial artery. See figures 9*a-b*, size of opening is 1.6.

The stapes in the odontocetes served as a support for the stapedial artery whose course passes through the stapedial opening, and as a support for the incus. The long crus of the incus terminates in its lenticular process which process attaches to a small depression in the head of the stapes. The short crus of the incus is connected to the fossa incudis by means of the posterior ligament of the incus. A tendon from the stapedius muscle is inserted into the posterior surface of the neck of the stapes. The muscle is supplied by the tympanic nerve, a branch of the facial nerve. The incus, in turn, articulates with the head of the malleus (figs. 8*a-d*). The manubrium of the malleus is attached to the tympanic membrane throughout its length, ending at the umbo at the center of the membrane. These details are given to indicate that the ossicles of the ear are interdependent. Although the stapedial artery has atrophied in some species of odontocetes, and hence has no need of support by the stapes, the stapes itself is still functioning as a support for the incus. It appears that the stapes attained its present state by Middle Miocene and has been relatively stable since. Its stability makes it of no particular use in determining relationships with other delphinids which lived since Middle Miocene time. In the mysticete whales the intercural opening remains large and functional, for example in *Balaenoptera physalus* (Fraser and Purves, 1960: pl. 53), in which not only the stapedial artery but also the internal carotid artery sometimes atrophy. The use of the opening in the stapes as a pasage for the stapedial artery has thus ceased in many species of whales. Slijper (1936:84) cites the work of several other authors who were unable to find a stapedial artery in the embryos of certain species of toothed whales. It is possible, however, that the small stapes is often lost during the cleaning of a fossil specimen, as it is lost frequently in the cleaning of skulls of Recent whales by compressed air or other means.

### CERVICAL VERTEBRAE

All of the cervicals are present and all have been distorted from their normal form by crushing, as have the adjoining dorsals. The striking character of the cervicals is the thinness of the third and seventh on their dorsal sides. Here the length of the vertebra is only one or two millimeters as contrasted to a ventral length on the same vertebra of several times that amount. Magnification shows a thin strip of matrix between adjoining cervicals; there was no fusion. In the crushing process, the dorsal side of the animal was much more affected than the ventral, and hence the measurements of the lengths of centra were taken on the ventral side.

MEASUREMENTS OF THE CERVICALS

| | 1 | 2* | 3 | 4 | 5 | 6 | 7 |
|---|---|---|---|---|---|---|---|
| Length of centrum | 20.0 | 19.5 | 5.0 | 4.0 | 3.5 | 7.0 | 5.0 |
| Height of neural spine | — | — | — | 35.0 | 44.0 | 45.0 | 43.0 |

* Odontoid process is missing.

MEASUREMENTS OF THE ATLAS (CERVICAL NO. 1)

| | |
|---|---|
| Length at base of neural spine | 20.0 |
| Height, excluding neural spine | 54.0 |
| Height of neural canal | 30.0 |
| Length of anterior cotyle | 32.0 |
| Width of anterior cotyle | 24.0 |
| Width across cranial side of anterior cotyles | 15.0 |
| Length of left transverse process | 13.0 |

*Dorsal vertebrae*—Thirteen additional vertebrae are scattered about in the matrix. All have transverse processes. There are also several ribs with head and tubercle and a sternal plate. These latter bones were left in the matrix because of their fragility.

MEASUREMENTS OF DORSALS ON VENTRAL SIDE

| | 1 | 2 | 3 | 4 | 5 | 6 | 7 | 8 | 9 | 10 |
|---|---|---|---|---|---|---|---|---|---|---|
| Length of centrum | 9.0 | 12.0 | 14.0 | 15.5 | — | — | — | — | 18.5 | 19.0 |
| Height of spine above centrum | 35.0 | 49.0 | 53.0 | 54.0 | 50.0 | — | — | — | — | — |

### LIMB BONES

The left forelimb is represented by a part of the scapula, part of the humerus, and the radius and ulna are complete and in place. The proximal end of the humerus is small, whereas the distal end is anteroposteriorly expanded. The radius is similarly expanded distally, and in these features the humerous and radius resemble *Phocaena*. The anterior and posterior margins of the scapula, as a result of its expansion, approximate the frontal plane of the animal, and in this feature also resemble *Phocaena* and other phocaenine genera. The expansion of these bones in *Loxolithax stocktoni* is not so pronounced as in living genera, however.

MEASUREMENTS OF THE HUMERUS

| | |
|---|---|
| Approximate length along midline of shaft | 69.0 |
| Anteroposterior dimension at distal end | 23.5 |
| Externointernal dimension at distal end | 17.0 |

MEASUREMENTS OF RADIUS AND ULNA

| | Radius | Ulna |
|---|---|---|
| Length along posterior border | 48.0 | 60.0 |
| Length along anterior border | 80.0 | 64.0 |
| Anteroposterior dimension, proximal end | 21.0 | 23.0 |
| Anteroposterior dimension, distal end | 28.5 | 18.5 |
| Anteroposterior dimension, narrowest part of shaft | 24.2 | 13.0 |
| Externointernal dimension, distal end | 11.0 | 8.5 |

Two fragmentary carpal bones are present but were not extracted from the matrix because of their preservation.

## COMPARISONS AND RELATIONSHIPS

The new species shows several significant similarities to *Phocaena vomerina*? Gill (Dall, 1874:298) (figs. 7*e-f*) in structure of the tympanic bulla, small size, lack of a projecting beak, and some teeth constricted below the crown and some expanded at the apex. The dental formula of *P. vomerina*? varies from 21/24 to 26/24. The new species exceeds the maximum with a maxillary count of 30 to 32? It also resembles *Delphinus bairdii*? (Dall, 1874:283) in the topography of the ventral face of the tympanic bulla (fig. 7), in small size of the teeth, and the inward curvature of their crowns. The dental formula of *D. bairdii*? Dall is 53/47, which removes it from consideration as a close relative of the new species. The comparison of the new species with the type of *Loxolithax* is detailed in the species diagnosis. Considering the nature of the dentition it should be mentioned that the living species *Stenella longirostis* (Cope) Dall (1874:288) has a dental formula 30/27 close to that of the new spcies, *L. stocktoni*.

For comparison of the skeletal parts of the new species with other genera, *Delphinodon dividum* True (1912) and *Kentriodon pernix* Kellogg (1927) have been selected because these three delphinids are approximately of the same size. *Delphinodon* and *Kentriodon* are represented by nearly complete skeletons. The geologic provenance of *Delphinodon* and *Kentriodon* is Middle Miocene, Calvert

Formation of Maryland, *Loxolithax sinuosa,* type species of the genus is from the Temblor Formation that on the evidence of diatoms (Lohman) is of the same age as the Calvert. The new species *(L. stocktoni)* is of Late Miocene age on the evidence from diatoms (Lohman) and foraminifera, thus it is younger than the genotypic, species *Delphinodon,* and *Kentriodon.*

| *Features used* | *Delphinodon dividum* | *Loxolithax stocktoni* | *Kentriodon pernix* |
|---|---|---|---|
| Vertex | small, pentagonal | small | small, pentagonal |
| Frontals | contact lambdoid crest at supraoccipital | fused; contact supraoccipital crest | large, pentagonal; contact supraoccipital |
| Nasals | large, depressed anteriorly | long, narrow, posteriorly thick | large, elevated and notched anteriorly |
| Maxillaries | not in contact with supraoccipital | contact supraoccipital | contact supraoccipital |
| Lambdoid crest | undeveloped | slightly developed | well developed |
| Maxillary foramina | anterior to maxillary notches | right, level with notch left, behind maxillary notch | posterior to antorbital notches |
| Maxillary notch | shallow | moderate | deep, narrow |
| Premaxillary eminence | posterior to antorbital notch | in front of nares | from posterior border of nares to premaxillary foramina |
| Premaxillary foramina | just anterior to maxillary notches | posterior to maxillary foramina | on a transverse line with maxillary foramina |
| Rostrum | short; attenuated | short; stout | long, slender |
| Location of posterior end of premaxilla | touches the nasals | at posterior border of nares | posterior border of nares |
| Teeth | 27 – 27 / 26 – 26 | 30 to 32 ? | 40 – 40 / 38 – 38 |
| size | largest 20 length × 5 (diam.) | 15.5 × 3.5 | 38.3 × 3.5 |
| premaxillary teeth | 3 | 0 | 8 |
| accessory cusps | on posterior teeth | occasionally on posterior teeth | no cusps nor carina |
| crowns | recurved | recurved | recurved |
| enamel of crowns | rugose on all teeth | some rugose crowns | smooth enamel |
| carina | both anterior and posterior sides of teeth | none | none |
| Posterior projection of supraorbital process of frontal | has posterior process long and slender | well developed | well developed |
| Symphysis | long, recurved | 1/4 length of mandible | 1/3 length of mandible |
| Age | Calvert Formation Middle Miocene | upper part of Valmonte Diatomite Late Miocene | Calvert Formation Middle Miocene |

In addition, there are important differences in the vertebrae and transverse processes in these three genera.

A comparison of the periotic and tympanic bones of the species *L. stocktoni, Delphinus bairdii?* Dall, and *Phocaena phocaena,* bring out significant differences.

*Delphinus bairdii?* Dall (figs. 6*a-b*), has a posterior process with its maximum breadth at its posterior termination where its width equals its length (11.2), a proportion unique compared to the periotics from the Sharktooth Hill bone bed mentioned above. The face of this process is more deeply concave, with more osseous ridges, and with a higher rim of bone encircling the process than that of the new species. The proportions of the two periotics are quite different: Breadth length index of the new species is 58.8; of *D. bairdii?* 69.0, the same as the living dolphin from Half Moon Bay, California. The *pars cochlearis* of this *Delphinus* is globular and lacks the wedge-shape of that of the new species of *Loxolithax.* The periotic of the type of the genus (figs. 6*c-d*) and that of the new species (figs. 6*e-f*) were compared in the Diagnosis of the species.

The periotic of *Phocaena phocaena* (figs. 6*g-h*) B. T. Waters collection, has characters which distinguish it from either of the periotic just compared. The differences in measurements can be put in tabular form:

| | *Loxolithax stocktoni* | *Phocaena phocaena* | *Delphinus bairdii?* |
|---|---|---|---|
| Greatest length (tip of anterior process to tip of posterior process) | 31.6 | 29.0 | 29.0 |
| Width in line with *fenestra ovalis* | 18.5 | 16.0 | 20.0 |
| Index — B/L | 58.8 | 55.2 | 69.0 |
| Distance between nearest edge of *fenestra rotunda* and tip of posterior process | 18.4 | 13.5 | 17.0 |
| Same to anterior process | 19.4 | 17.2 | 18.0 |
| Greatest length of posterior process | 14.5 | 17.0 | 11.2 |
| Greatest width of posterior process | 10.9 | 9.3 | 11.2 |

L = the straight-line distance from the tip of the posterior process to the tip of the anterior process.

B = breadth of periotic = distance, external face of periotic to the internal face of the *pars cochlearis* measured in line with the stapes.

Index = $\frac{B \times 100.}{L}$

The measurements accent the narrowness of the *Phocaena* periotic; the closeness of the *fenestra rotunda* to the posterior process, and the narrowness of the posterior process. The face of the posterior process of the *Phocaena* is arched with the high area being a longitudinal ridge from which the slope of the face is toward the edges. The face is otherwise free of osseous ridges, in contrast with the other two periotics.

In 1931 Kellogg studied 24 periotic bones of small delphinids from a bed three feet thick in the Temblor Formation at Sharktooth Hill, near Bakerfield, California. Of these bones 14 belonged to the California Academy of Sciences, San Francisco, and 10 belonged to the United States National Museum in Washington, D.C. Kellogg established seven genera and eight species, on the basis of these bones. The collection had eight right periotics, and 16 left. Of the specimens sufficiently complete for measurement for the following table, eight were right, and 11 left. The following table of measurements shows the comparative proportions of the periotics just mentioned, plus two Recent dolphins, a new species of *Loxolithax, Kentriodon pernix,* and *Delphinodon dividum.*

| *Taxon* | *Specimen number* | *Breadth* | *Length* | *Index* |
|---|---|---|---|---|
| DELPHINIDAE: | | | | |
| *Loxolithax sinuosa* | CAS 4352, type . . . . . . . | 16.8 | 29.6 | 56.8 |
| | CAS 4351, paratype . . . . . | 16.3 | 28.0 | 58.2 |
| *Loxolithax stocktoni* | UCMP 34576, type . . . . . . | 18.5 | 31.6 | 58.6 |
| *Kentriodon pernix* | USNM 8060, type . . . . . . | 16.9 | 28.8 | 58.7 |
| *Delphinodon dividum* | USNM 7278 . . . . . . . . | 18.0 | 30.0 | 60.0 |
| *Lagenorhynchus obliquidens* | UCMVZ 138675, right . . . . | 21.0 | 31.0 | 68.0 |
| | UCMVZ 138675, left . . . . . | 21.0 | 33.0 | 64.0 |
| *Delphinus delphis* | USNM 21525 . . . . . . . . | 20.0 | 30.0 | 66.7 |
| *Grypolithax obscura* | CAS 4349, type . . . . . . . | 16.4 | 27.1 | 60.5 |
| | CAS 4347, paratype . . . . . | 17.2 | 28.7 | 59.7 |
| *Grypolithax pavida* | CAS 4348, type . . . . . . . | 16.2 | 26.9 | 60.0 |
| | CAS 4350, paratype . . . . . | 16.4 | 26.5 | 61.9 |
| | USNM 11575, paratype . . . . | 16.5 | 26.6 | 62.0 |
| *Platylithax robusta* | CAS 4339, type . . . . . . . | 18.9 | 30.0 | 63.0 |
| *Nannolithax gracilis* | USNM 11569, type . . . . . | 14.2 | 23.8 | 59.7 |
| *Lamprolithax annectens* | CAS 4343, type . . . . . . . | 18.0 | 27.9 | 64.5 |
| *Lamprolithax simulans* | USNM 11566, type . . . . . | 16.4 | 26.3 | 62.4 |
| | USNM 11567, paratype . . . . | 16.1 | 26.3 | 61.2 |
| | CAS 4344 . . . . . . . . . | 16.3 | 25.5 | 64.0 |
| | CAS 4345 . . . . . . . . . | 15.9 | 26.7 | 59.6 |
| *OEdolithax mira* | USNM 11572, type . . . . . | 17.6 | 29.6 | 59.5 |
| | USNM 11571, paratype . . . . | 17.2 | 26.4 | 65.2 |
| *Liolithax kernensis* | USNM 10854 . . . . . . . . | 18.2 | 29.5 | 61.8 |
| | CAS 4340, type . . . . . . . | 20.2 | 31.8 | 63.6 |
| | USNM 11565 . . . . . . . . | 19.7 | 30.8 | 64.0 |
| *Kampholophos serrulus* | UCMP 36045, left . . . . . | 18.8 | 32.2 | 58.5 |
| | UCMP 36045, right . . . . . | 18.4 | 33.9 | 54.4 |
| PHYSETERIDAE: | | | | |
| *Aulophyseter morricei* | USNM 11230, type . . . . . | 25.2 | 38.1 | 66.2 |
| | USNM 10853 . . . . . . . | .26.0 | 41.0 | 63.5 |
| | CAS 2798 . . . . . . . . . | 25.1 | 37.6 | 66.7 |
| | CAS 2705 . . . . . . . . . | 28.5 | 42.0 | 68.0 |
| | CAS 2794 . . . . . . . . . | 25.0 | 36.2 | 69.1 |
| | CAS 2800 . . . . . . . . . | 27.0 | 42.5 | 63.5 |
| | CAS 2801 . . . . . . . . . | 23.1 | 36.7 | 63.0 |
| SQUALODONT: | | | | |
| *Squalodon calvertensis* | NSNM 10484 . . . . . . . . | 21.3 | 41.5 | 51.4 |
| ARCHAEOCETE: | | | | |
| *Zygorhiza kochii* | USNM 10855 (immature) . . . | 26.5 | 64.0 | 41.5 |
| | USNM 11962 (right) . . . . | 27.0 | 75.0 | 36.0 |
| | USNM 11962 (left) . . . . . | 26.8 | 78.0 | 34.4 |

SUMMARY

| *Taxa* | *Range of index* |
|---|---|
| Recent delphinids . . . . . . . . . | 66.7–68.0 |
| Twenty-five fossil delphinids . . . . . . | 56.2–65.6 |
| *Aulophyseter* . . . . . . . . . . . . | 63.0–69.0 |
| *Squalodon* . . . . . . . . . . . . . | 51.4 |
| *Zygorhiza* (adult only) . . . . . . . . | 34.4–36.0 |

The above 38 periotics fall into several groups:

1. Twenty periotics of fossil delphinids from the Temblor. *Nannolithax,* an extreme form, is omitted (probably it is a phocaenid).

ACTUAL MEASUREMENTS

| | *Liolithax* | *Lamprolithax simulans* | | (Next smallest after *Nannolithax*) |
|---|---|---|---|---|
| B | 19.7 | 15.9 | 3.8 | 3.8/15.9 = 23.8% variation in B in 19 periotics |
| L | 30.8 | 26.7 | 4.1 | 4.1/26.7 = 15.7% variation in L in 19 periotics |
| | 64.0 | 59.6 | 4.4 | = 4.4% variation in B/L ratio |

In this group the width varies more than the length.

2. Sperm whales: 11 periotics were found; 7 of them measurable for this table; all are of the same species, all from a three-foot bed in the Temblor.

Measured variation in breadth = 5.4, and 5.4/23.1 (smallest) = 23.4% variation in B
Measured variation in length = 6.3, and 6.3/36.2 (smallest) = 17.3% variation in L

In this group the width is more variable than the length. Kellogg (1931:366) states "It is remarkable how closely, except for minor modifications in the anterior process, these periotics resemble those of recent sperm whales, for the relative proportions and peculiarities are essentially the same." The index numbers of these seven periotics partly overlap those of the recent dolphins. The sperm attained his modern ratio of breadth/length by Middle Miocene time.

3. Only one squalodont of Middle Miocene is available for measurement at this time. Its index number is 51.4; it may be interesting to note that the midpoint between the Recent dolphin (68.0) and the measured archaeocete (34.4) is 52.0.

4. The immature specimen of *Zygorhiza* has already attained adult width, but not length so the index number is too high. The overall picture from these few samples is that from Archaeocete to Recent dolphin, the breadth of the periotic has been increasing with respect to the length. The increase in size is in the *pars cochlearis,* more specifically in the osseous labyrinth; vestibule, semicircular canals, and cochlea, which comprise the essential parts of the mechanics of hearing. The range of index numbers shows an enlarging cochlea from the primitive Archaeocete to *Squalodon* to Middle Miocene delphinids to Sperms to modern dolphin, with the Sperms overlapping Miocene delphinids and some modern dolphins. Variation of the length of the periotics (*Kampholophos* Rensberger, 1969) is 5.27% in the same individual; and in the modern dolphin from Half Moon Bay 4.55%.

Examination of the surface outline of the *fenestra rotunda* in a collection of ear bones from a three-foot thick bed in the Temblor Formation of Sharktooth Hill (page numbers in Kellogg [1931] are given) can be summarized as follows:

| | Page | Surface outline of *fenestra rotunda* |
|---|---|---|
| *Liolithax kernensis* | 376 | circular |
| *OEdolithax mira* | 379 | crescentic |
| *Lamprolithax annectens* | 384 | kidney-shaped–reniform |
| *Nannolithax gracilis* | 387 | circular |
| *Platylithax robusta* | 389 | hemi-circular |
| *Loxolithax sinuosa* | 380 | triangular |
| *Grypolithax pavida* | 397 | crescentic |

| | | |
|---|---|---|
| *Loxolithax stocktoni* | Late Miocene | spherical triangle |
| *Kenriodon pernix* | Calvert Miocene - Middle | subtriangular |
| Recent porpoise | Half Moon Bay, California | spherical triangle |

This sample suggests that the surface outline of the *fenestra rotunda* has no significance as an aid to determining the genus of the delphinid.

## SUMMARY

This paper contributes information on the skull and some other skeletal parts of *Loxolithax*, a genus of delphinid, originally based on a single left periotic bone (type) supported by another left periotic (paratype) from the Middle Miocene Temblor Formation near Bakersfield, California.

The delphinid described in this paper came from the upper part of the Valmonte Diatomite Member of the Monterey Shale and is Late Miocene in age. The locality is in the Palos Verdes Hills, about 20 miles south of Los Angeles, California. The delphinid of this paper is a new species of the genus *Loxolithax*.

The evidence from the species described herein is that the stapes has changed so little between Upper Miocene time and the present, that it has no specific nor generic diagnostic value in the relationships of delphinids.

Breadth/Length index numbers are calculated to show the variation in 24 periotic bones on which Kellogg established seven genera and eight species. The index numbers were calculated for other cetaceans so that a total of 38 periotics from Eocene to Recent cetaceans are considered. The general directional change in the periotic over this long period of time has been an increasing of width with respect to the length. The increased width was the result of the increased size of the *pars cochlearis* which houses the vestibule, semicircular canals, and cochlea, the essential structures of the mechanics of hearing. It is suggested that index numbers of this kind be calculated for Paleocene and Eocene land mammals in order to learn which, if any of them, have proportional ratios similar to those of the Archaeocetes.

# LITERATURE CITED

ABEL, O.

1905. Les Odontocètes du Boldérien (Miocene Supérieur) D'Anvers. Extrait des Mémoires du Musée Royal D'Histoire Naturelle de Belgique, 3:139 et seq.

BRANDT, J. F.

1871. Bericht über den Fortgang meiner Studien über die Cetaceen, welche das grosse zur Tertiärzeit von Mitteleuropa bis Centralasien hinein ausgedehnte Meeresbecken bevölkerten. Mélanges biol. tirés de Bull. acad. imp. St. Pétersbourg, 8:193. Vergl. auch: Bull., 16:566, Oct.

CASE, E. C.

1904. Miocene Text, Maryland Geol. Surv., Baltimore, pp. 12–13, pl. 11.

CLAUDIUS, M.

1858. Physiol. Bemerk. über das Gehörorgan der Cetaceen und das Labyrinth der Säugetiere. Kiel.

COPE, E. D.

1868. Proc. Acad. Nat. Sci. Philadelphia 19:132, 145, Mar. 10, 1868 and Jour. Acad. Nat. Sci. Philadelphia, ser. 2, 6:91, 1868.

DALL, W. H.

1874. Catalogue of the Cetacea of the North Pacific Ocean; pp. 291–308. (Appendix to: The Marine Mammals of the Northwestern Coast of North America by Scammon, C. M., repr. 1968.)

DENKER, A.

1902. Zur Anatomie des Gehörorgans der Cetacea. Anat. Hefte, 19:423–445.

DORAN, A. H. G.

1879. On the morphology of the mammalian ossicula auditus. Trans. Linn. Soc. London, 2d. ser., 1 (Zool.):371–497, pls. 58–64, 258 figs.

DUBUS, B.

1867. Sur quelques Mammiféres du Crag d'Anvers. Bull. Acad. Roy. Sci. Belgique, ser. 2, 24:569.

1872. Mammiféres nouveaux du Crag d'Anvers, Bull. Acad. Roy. Soc. Belgique, Bruxelles, ser. 2, 34:492.

FRASER, A.

1882. On the development of the ossicula auditus in the higher Mammalia. (Abstr.) Proc. Roy. Soc. London, 33:446. (Communicated by A. Thompson.)

1883. On the development of the ossicula auditus in the higher Mammalia. Philos. Trans. Roy. Soc. London, 173 (pt. iii):901–925, pls. 54–58, 40 figs.

FRASER, F. C., and P. E. PURVES

1960. Hearing in Cetaceans. Bull. Brit. Mus. (Nat. Hist.), vol. 7, no. 1 of Zool. Ser.

GOODRICH, E. C.

1930. Studies on the structuer and development of vertebrates, 1:449–485. (1958 Dover Publ. Inc., N.Y., reprint.)

HARLAN, R.

1824. Jour. Acad. Nat. Sci. Philadelphia, ser. 2, 4:232, pl. 14, fig. 1.

HYRTL, J.

1845. Vergleichend-Anatomische untersuchungen über das Innere Gehörorgan des Menchen und des Säugetiere., 7:139, 9 pls. Prague.

KELLOGG, A. R.

1925. On the occurrence of remains of fossil porpoises of the genus *Eurhinodelphis* in North America. Proc. U.S. Nat. Mus., 66 (publ. 2563):1–40, pls. 1–17, text figs. 1–4.

1927. *Kentriodon pernix*, a Miocene porpoise from Maryland. Proc. U.S. Nation. Mus., 69 (art. 19):1–55, pls. 1–14.

1931. Pelagic mammals from the Temblor Formation of the Kern River Region, California. Proc. Calif. Acad. Sci., 4th ser., 19(12):217–397.

1936. A review of the Archaeoceti. Carnegie Inst. Washington Publ., no. 482, pp. 1–357; pls. 1–37; 88 figs. in text.

LEIDY, J.

1851. Proc. Acad. Nat. Sci. Philadelphia, 5:326–327.

MAKIYAMA, J.

1936. *Sinanodelphis izumidaensis:* a new Miocene dolphin of Japan. Mem. Coll. Sci., Kyoto Imp. Univ., ser. B, 11(2, art. 3):115–134, 2 figs., pls. I–III.

RENSBERGER, J. M.

1969. A new iniid cetacean from the Miocene of California, Univ. Calif. Publ. Geol. Sci., 82: 1–33, pls. 1–4, 2 figs. in text.

SLIJPER, E. J.

1936. Die. Cetaceen. Vergleichend-Anatomisch und Systematisch. Cap. Zool., 7(1):1–304; 7(2):305–590. M. Nijhoff. The Hague.

TRUE, F. W.

1908. On the occurrence of remains of fossil cetaceans of the genus *Schizodelphis* in the United States, and on *Priscodelphinus* (?) *crassangulum* Case. Smithsonian Misc. Coll., 50 (publ. 1782):449–460, pls. 59–60.

1912. Description of a new fossil porpoise of the genus *Delphinodon* from the Miocene formation of Maryland. Jour. Acad. Natur. Sci., Philadelphia, ser. 2, 15:165–194, pls. 17–26.

WOODRING, W. P., M. N. BRAMLETTE, and W. S. W. KEW

1946. Geology and Paleontology of Palos Verdes Hills, California. U.S. Geol. Surv. Prof. Paper. 207.

# FIGURES

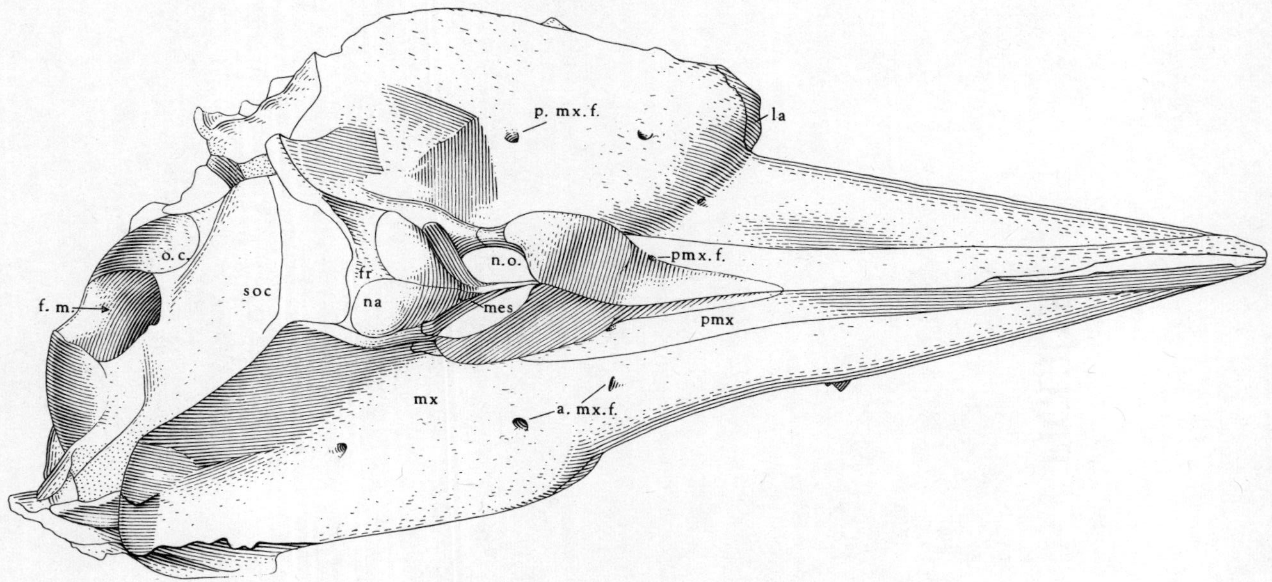


Fig. 1. *Loxolithax stocktoni*, n. sp., type skull, U.C.M.P. 34576, dorsal view, × 1/2. Abbreviations: a. mx. f.—anterior maxillary foramina, al—alisphenoid, bs—basisphenoid, boc—basioccipital, e. a. m.—external auditory meatus, eoc—exoccipital, f. m.—foramen magnum, f. p.—falcate process of basioccipital, fr—frontal, g. f.—glenoid fossa, ju—jugal, la—lacrimal, m. n.—maxillary notch, mes—mesethmoid, mx—maxilla, na—nasal, n. o.—narial opening, o. c.—occipital condyle, o. g.—optic groove, os—orbitosphenoid, p. mx. f.—posterior maxillary foramen, p. p.—paroccipital process, pa—parietal, pal—palatine, po. p.—postorbital process of frontal, pmx—premaxilla, pmx. f.—premaxillary foramen, pt—pterygoid, soc—supraoccipital, s. p.—styliform process of jugal, and sq—squamosal.

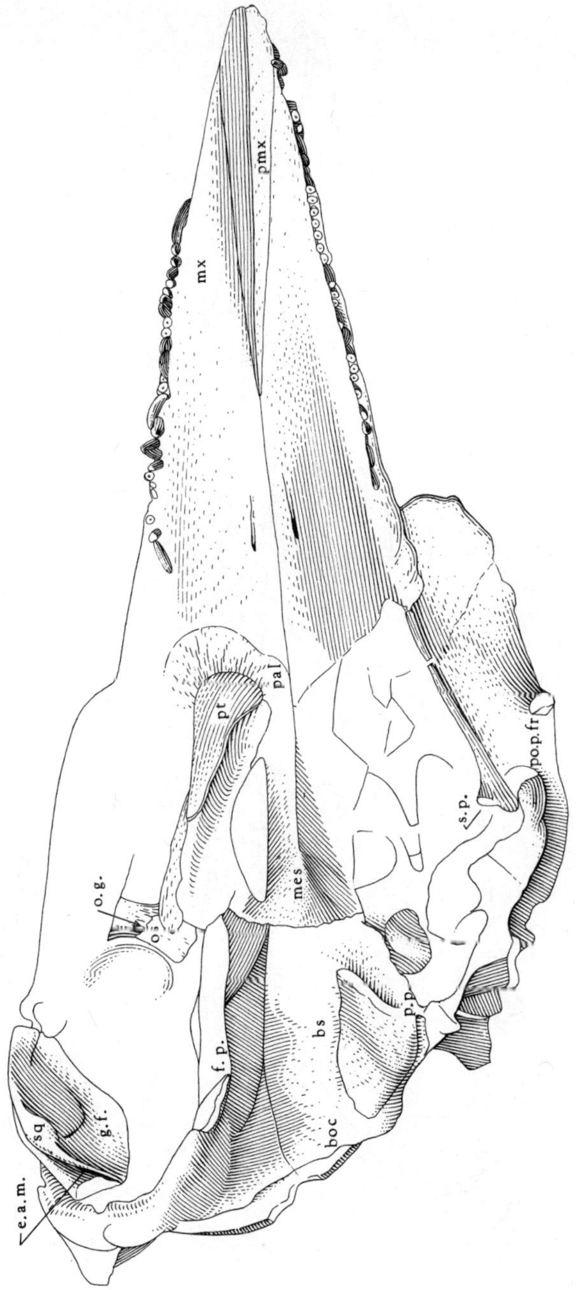


Fig. 2. *Loxolithax stocktoni*, n. sp., type skull, U.C.M.P. 34576, ventral view, × 1/2. The posterior parts of both lower jaws are crushed into the zygomatic-orbital areas, obscuring most of the features in those areas. (See fig. 1 for abbreviations.)

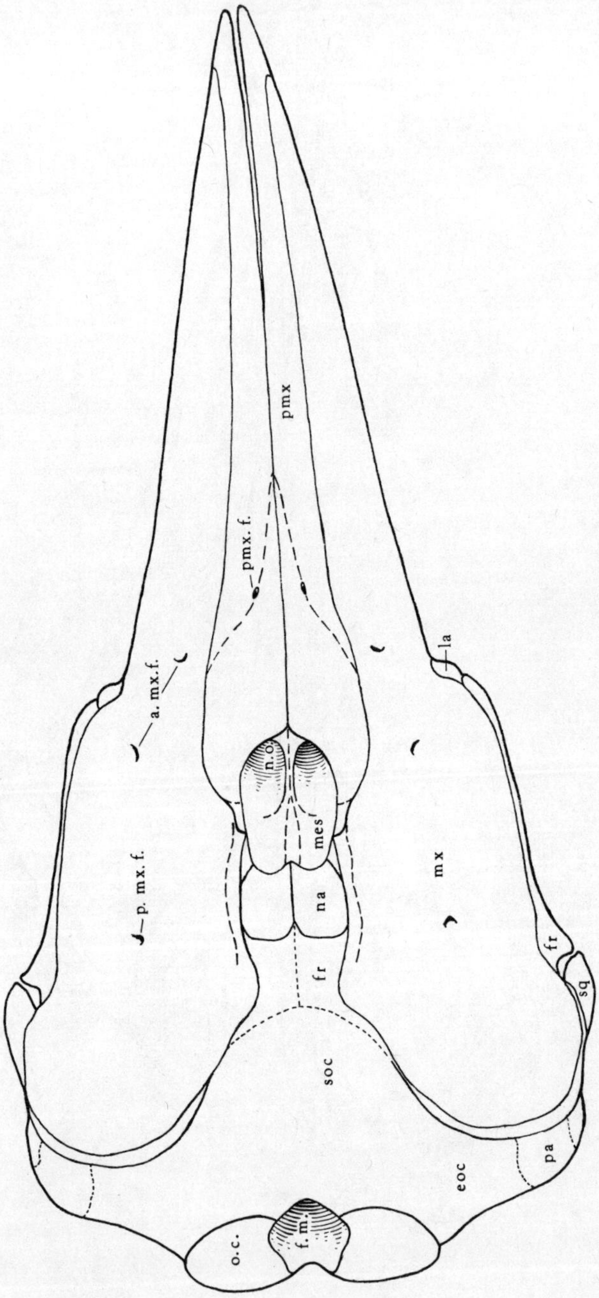


Fig. 3. ***Loxolithax stocktoni***, n. sp., restoration of skull of U.C.M.P. 34576, dorsal view, × 1/2. (See fig. 1 for abbreviations.)

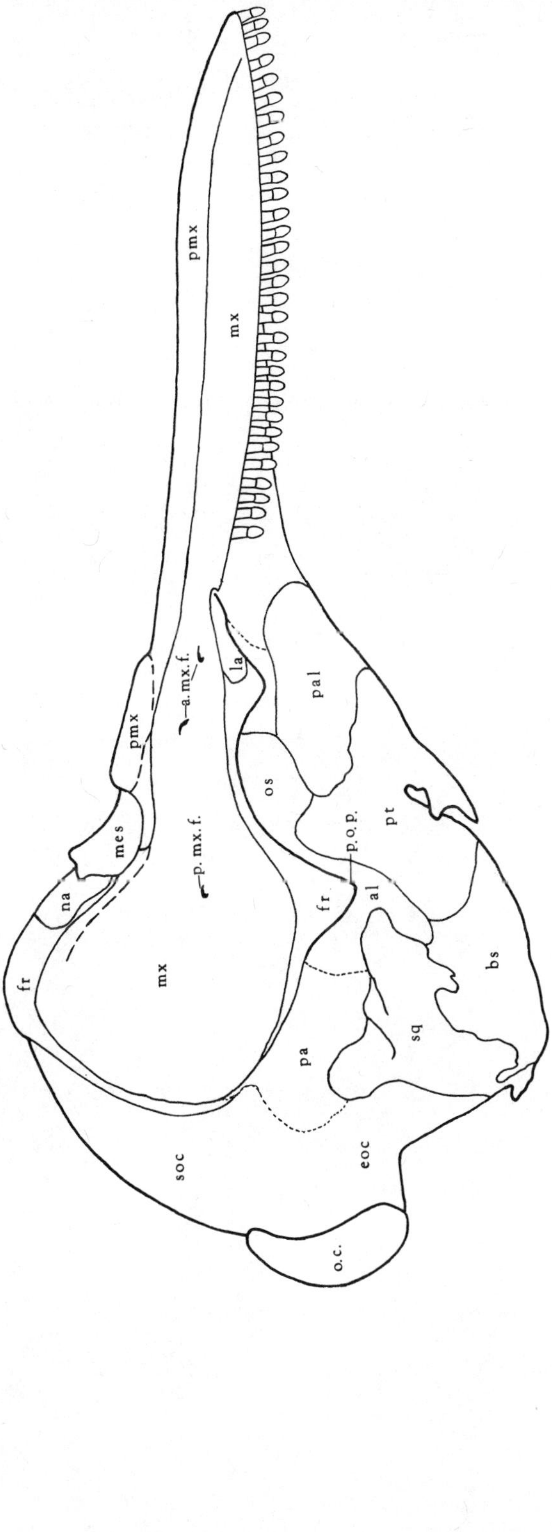


Fig. 4. *Loxolithax stocktoni*, n. sp., restoration of skull of U.C.M.P. 34576, lateral view, × 1/2. (See fig. 1 for abbreviations.)

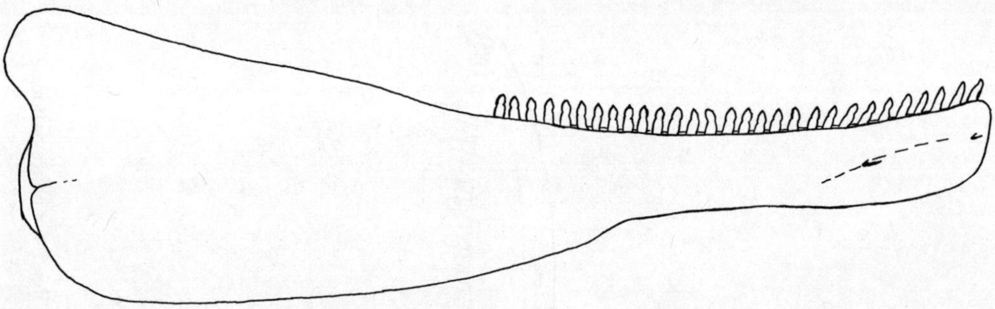

Fig. 5. *Loxolithax stocktoni,* n. sp., restoration of right lower jaw of U.C.M.P. 34576, mesial view, × 1/2.

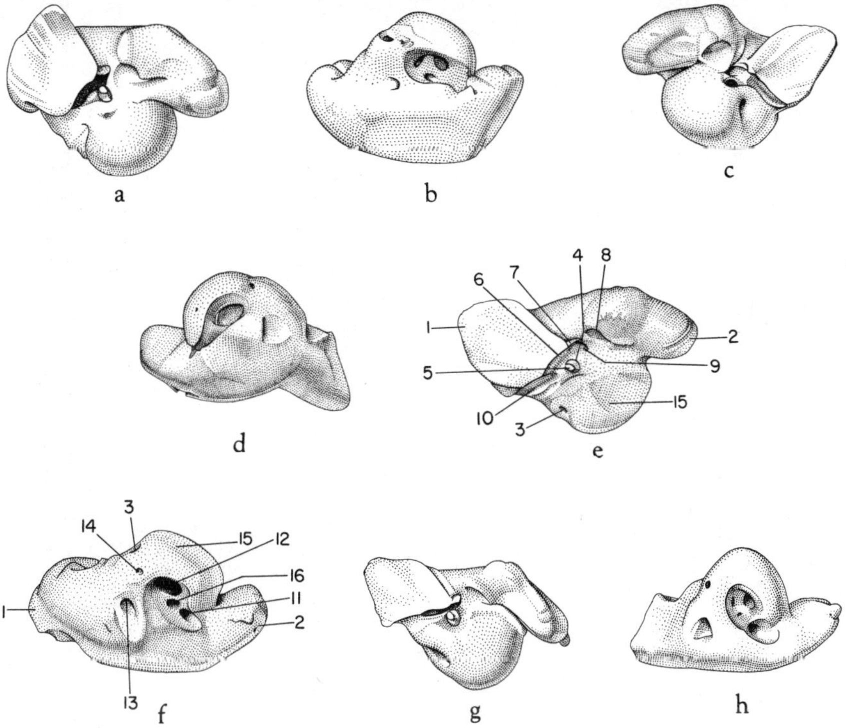


Fig. 6. (All drawings are × 1.) *a. Delphinus bairdii* Dall, J. H. Hutchison Collection, ventral (or epitympanic) face of right periotic. *b. Delphinus bairdii* Dall, J. H. Hutchison Collection, dorsal (or cerebral) face of right periotic. *c. Loxolithax sinuosa* Kellogg, type, C.A.S. 4352, ventral face of left periotic. *d. Loxolithax sinuosa* Kellogg, type, C.A.S. 4352, dorsal face of left periotic. *e. Loxolithax stocktoni,* n. sp., U.C.M.P. 34576, ventral face of right periotic. *f. Loxolithax stocktoni,* n. sp., U.C.M.P. 34576, dorsal face of right periotic. *g. Phocaena phocaena* Gray, B. T. Waters Collection, ventral face of right periotic. *h. Phocaena phocaena* Gray, B. T. Waters Collection, dorsal face of right periotic.

Legend:

1–Posterior process
2–Anterior process
3–*Fenestra rotunda*
4–*Fenestra ovalis* at foot of stapes on a, e and g)
5–Stapes
6–Canal for facial nerve
7–Fossa for incus
8–Fossa for head of malleus
9–Ventral (epitympanic) orifice of facial canal
10–Fossa for stapedial muscle
11–Dorsal (cerebral) orifice of facial canal
12–Spiral tract, ending in *foramen centrale*
13–Dorsal orifice of *aqueductus vestibuli*
14–Dorsal orifice of *aqueductus cochleae*
15–*Pars cochlearis*
16–*Foramen singulare*

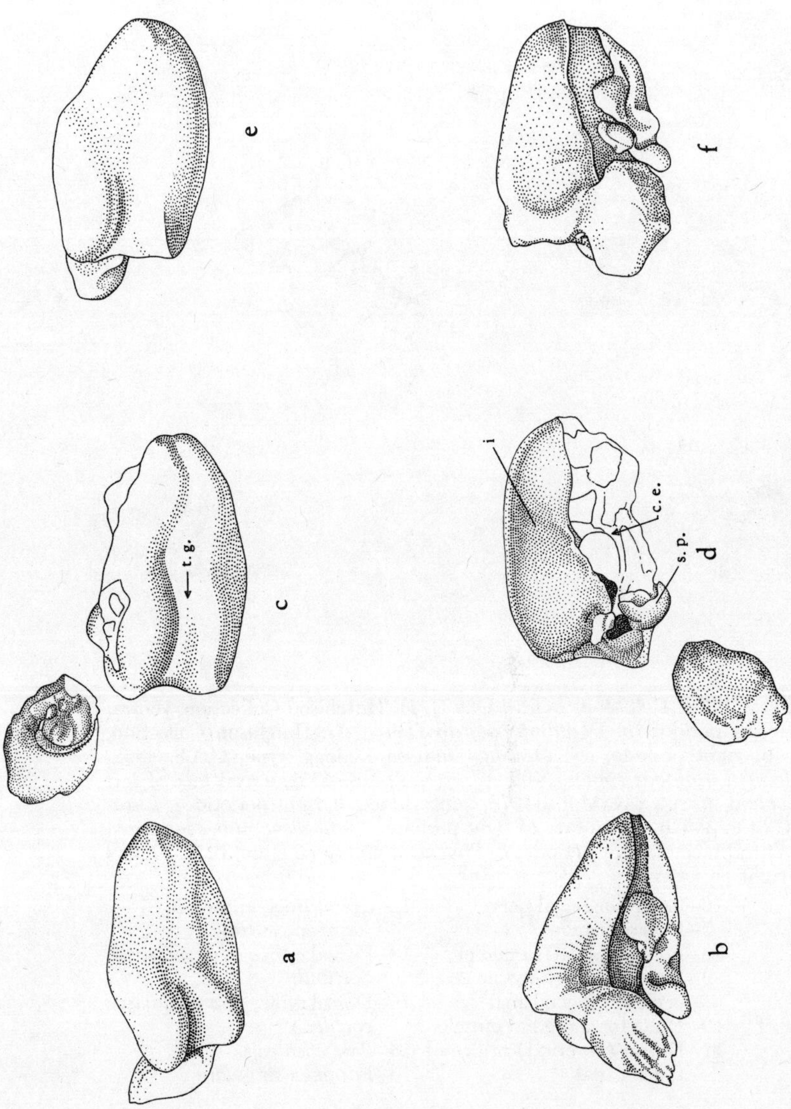


Fig. 7. (All drawings are × 1.) *a. Delphinus bairdii?* Dall, J. H. Hutchison Collection, ventral face of right tympanic bulla. *b. Delphinus bairdii?* Dall, J. H. Hutchison Collection, dorsal face of right tympanic bulla. *c. Loxolithax stocktoni*, n. sp., U.C.M.P. 34576, ventral face of right tympanic bulla; posterior pedicle is on left. *d. Loxolithax stocktoni*, n. sp., U.C.M.P. 34576, dorsal face of right tympanic bulla; posterior pedicle is on left. *e. Phocaena phocaena* Gray, B. T. Waters Collection, ventral face of right tympanic bulla. *f. Phocaena phocaena* Gray, B. T. Waters Collection, dorsal face of right tympanic bulla.

Legend:
c. e.—*cavum epitympanicum*
i —*involucrum*
s. p.—sigmoid process
t. g.—tympanic groove

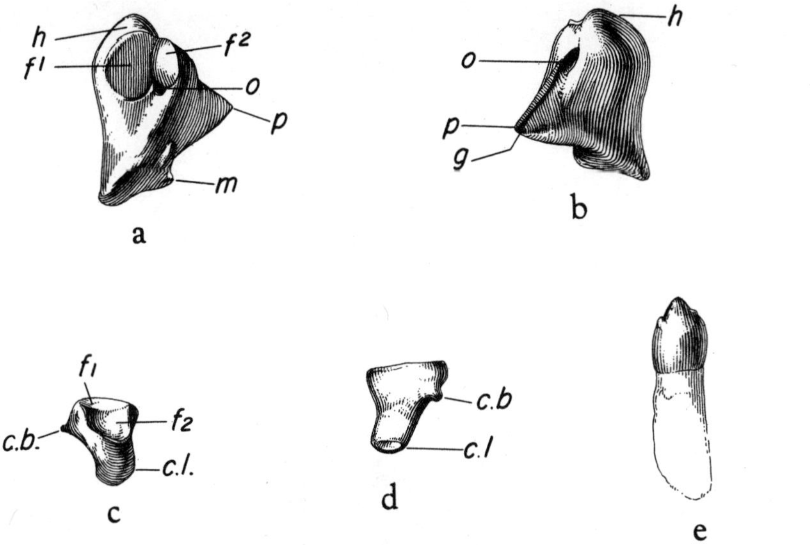


Fig. 8. *a. Loxolithax stocktoni,* n. sp., U.C.M.P. 34576, posterointernoventral face of right malleus, × 4. *b. Loxolithax stockoni,* n. sp., U.C.M.P. 34576, anteroexternal face of right malleus, ×4. *c. Loxolithax stocktoni,* n. sp., U.C.M.P. 34576, anteroexetrnal face of right incus, × 4. *d. Loxolithax stocktoni,* n. sp., U.C.M.P. 34576, ventral face of right incus, × 4. *e. Loxolithax stocktoni,* n. sp., U.C.M.P. 34576, a tooth with accessory cusps, × 3.

Legend:

c. b. —*crus breve*
c. l. —*crus longum*
$f^1$, $f^2$—facets for articulation with incus
$f_1$, $f_2$—facets for articulation with head of malleus (fitting against $f^1$, $f^2$)
g —groove along anterior process for *chorda tympani* nerve
h —head
m —manubrium
o —opening = *Fovea lateralis* of Pompeckj for *chorda tympani* nerve
p —anterior process of malleus

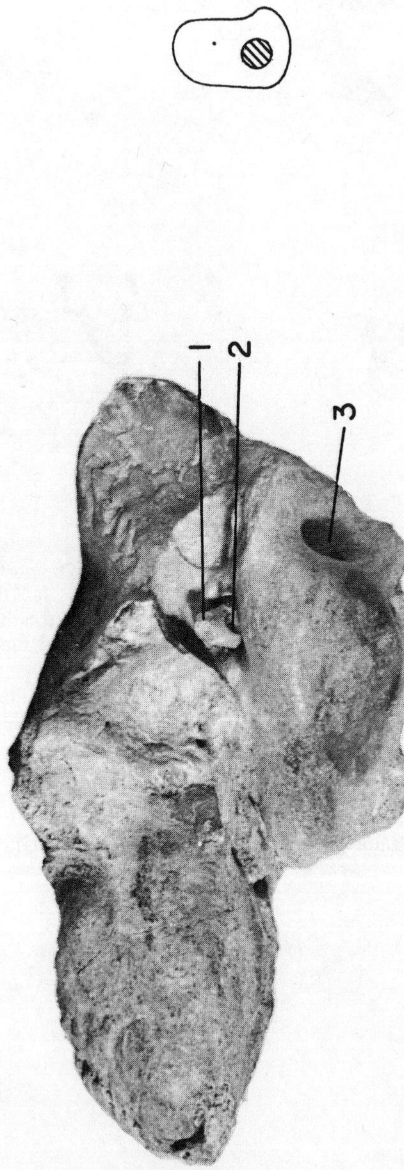


Fig. 9. *Kekenodon onamata* Hector, type Dominican Museum of New Zealand No. 306. *a*. Ventral face of left periotic. (1) Stapes; (2) Opening for stapedial artery; (3) *Fenestra rotunda*. *b*. Mesial view of stapes showing stapedial foramen. Diameter of the aperture is 1.6 mm.